José Maria Filippini Alba

Geochemical mapping in Uruguay

José Maria Filippini Alba

Geochemical mapping in Uruguay

Based on data from current sediments and quartz veins in Precambrian terrain

ScienciaScripts

Imprint
Any brand names and product names mentioned in this book are subject to trademark, brand or patent protection and are trademarks or registered trademarks of their respective holders. The use of brand names, product names, common names, trade names, product descriptions etc. even without a particular marking in this work is in no way to be construed to mean that such names may be regarded as unrestricted in respect of trademark and brand protection legislation and could thus be used by anyone.

Cover image: www.ingimage.com

This book is a translation from the original published under ISBN 978-3-330-77085-0.

Publisher:
Sciencia Scripts
is a trademark of
Dodo Books Indian Ocean Ltd. and OmniScriptum S.R.L publishing group

120 High Road, East Finchley, London, N2 9ED, United Kingdom
Str. Armeneasca 28/1, office 1, Chisinau MD-2012, Republic of Moldova, Europe
Managing Directors: Ieva Konstantinova, Victoria Ursu
info@omniscriptum.com

Printed at: see last page
ISBN: 978-620-8-63388-2

Summary

Presentation

This book presents geochemical data from Uruguay collected in the 1980s and reprocessed in the 1990s on the individual initiative of the author, who was an employee of the Divisiόn Nacional Mineria y Geologia (DINAMIGE), a mixture of the secretariat of mines and the geological service, until 1996.

Four institutions were fundamental in the process, the State University of Sao Paulo (Brazil) where the author did his master's degree in geochemistry (1989 - 1993) and later his doctorate in geochemistry (1994 - 1998), with collaboration from the State University of Campinas (Brazil), the Geological Survey of Finland and the Academic University of Angus (Finland) in the last period.

Stream sediment geochemistry was widely used in the context of mineral prospecting until the early 1980s, when it took on an environmental orientation, mainly to characterize the heavy metal content of various geomaterials, but also with a focus on geochemical mapping. The geochemistry of quartz veins, on the other hand, is almost entirely focused on the mineral sector, involving prospecting and exploration.

The text mentions geochemical studies in two areas located in the Uruguayan crystalline, one on the outskirts of the town of *Isla Patrulla* and the other involving four 1:50,000 sheets from Uruguay's national cartographic plan: *Gutiérrez, José P. Varela, Pirarajà* and Zapicà. *Varela, Pirarajà* and *Zapicàn*. The main results of the above-mentioned master's and doctoral theses were presented using statistics, graphs and maps in an attempt to interpret the geochemical behavior of the various data sets, establishing relationships with geological and mineral aspects in a regional or local context.

The geochemical response of current sediments showed a differentiation in the age of the geological formations, from the lower-middle Proterozoic on the one hand and the upper Proterozoic on the other. The local anomalies were mainly associated with the *Lavalleja* group, and two groups with different behaviour were characterized around *Isla Patrulla*, one related to Fe and the other to Fe-Mn. The geochemistry of the quartz veins was controlled by accessory minerals (oxides, sulphides...). Both sets of data contained samples related to mafic-ultramafic rocks.

Preface

When I began my studies in chemistry at the Universidad de la Republica (UdelaR, Uruguay) in the early 1980s, I was unfamiliar with the principles of geology and geochemistry, which I encountered when I joined the Department of Mining and Geology (DINAMIGE, Uruguay) three years later as a technical assistant. It was a regional geochemical prospecting program in partnership with the Institute of Geological and Mining Research (BRGM, France).

With the political transition that took place in Uruguay in 1984, there were changes of direction in the mineral sector, and a cooperation agreement was initiated with the Federal Institute of Geology and Natural Resources (BGR, Germany) and the aforementioned program was abolished in 1988.

At the time, DINAMIGE had no geoscience specialists, and the technical-professional positions were filled by agronomists, surveyors, geology graduates, chemists and students. So the BGR, in partnership with the German Academic Exchange Agency (DAAD), decided to offer undergraduate and postgraduate scholarships to Uruguayan citizens through two calls for proposals, one in 1986 and the other in 1989. For various reasons, in the second call for proposals I was awarded a scholarship for postgraduate studies in geochemistry, which had to be carried out in a regional context. The State University of Sao Paulo (USP, Brazil) was the chosen option and I started my master's degree in "Geochemistry of exogenous processes" in August 1989. At the time, the geochemistry of current sediments was considered, which was compared to the geochemistry of quartz veins using statistical methods in the *Isla Patrulla* region. At the beginning of this period, there was significant support from Dr. Gernot Reitmayr (Head of Mission), Dr. Klaus Fesefeldt (Consultant) and Dr. Werner Adelhart (Exploration Geologist), now all retired from the BGR, who made the quartz vein data available and collaborated with the data processing.

After my master's degree, I rejoined DINAMIGE for two years, where I worked on the environmental assessment of mining projects in Uruguay. But I started my doctorate in 1994, at the invitation of Dr. Sônia Maria Barros de Oliveira (USP, Brazil), who had already been my master's supervisor. Dr. Sônia made a significant contribution to both theses, thanks to her accurate reasoning, her excellent writing skills and her qualification in the geochemistry of exogenous processes. As the subject of his thesis, he chose to compare geochemical data of current sediments with Landsat - TM orbital images in the *Gutiérrez, José P. Varela, Piraraja and Piraraja* regions. Although there was a good fit in the global context, on a local scale the geochemical response of current sediments had a weak correlation with the reflectance data. However, the integration of data into a geographic information system showed advantages when processing geospatial data using multi-criteria procedures.

This project was supported by Dr. Alvaro P. Crosta from the State University of Campinas

(UNICAMP, Brazil), who co-supervised the doctoral thesis, given his expertise in orbital data processing. He also made the georeferenced image processing laboratory available and contributed to the processing of the orbital data and the acquisition of the images.

As part of the doctoral project, and thanks to Dr. Sônia's contact with Dr. Alf Bjorklund of the Academic University of Âbo (UAÂ, Finland) and with geologist Toni Eerola, who worked in the international division of the Geological Survey of Finland (GTK), it was possible to spend six months at these institutions, with the contribution of researcher Nils Gustavsson (GTK) also standing out. Unfortunately, in the meantime I left DINAMIGE, but the geochemical data was processed until the end of my doctorate in 1998; five technical and scientific publications were generated, one in Spanish, two in Portuguese and two in English.

The way in which the data was processed, as presented in the text, has a peculiar quantitative profile, with significant use of statistical methods, including multivariate methods, such as principal component analysis, discriminant analysis and regression analysis, which allows an analogous perspective for the *background*, as well as the establishment of multiple associations that are associated with specific mineralogical or geological processes. This type of analysis is essential when evaluating potentially harmful *environmental substances* (PHES) to determine thresholds and averages in water, sediment, soil and plant samples.

Geographic *information* systems have facilitated the geospatial analysis of information, also known as GIS. The first time I heard of GIS was in 1993 with Idrisi®, and I later broadened the spectrum of its use with the U.S. Navy's GRASS software during my doctorate. During this period we also used the Australian software ERMapper® , which has some functions that resemble a GIS, but was developed for digital image processing. After completing my postgraduate studies, I started working with geotechnologies, where the use of GIS and digital processing of orbital images became commonplace. Other widely used software in Brazil is ENVI® for image processing and the ARC line® for GIS applications.

Geochemistry is a discipline that mainly integrates aspects related to chemistry and geology, but in Brazil it is only considered in the context of geology courses. From the perspective of chemistry, we would have associated concepts of mineralogy and inorganic and organic chemistry, with a broad field of research related to kerogen (the precursor of petroleum), organometallic complexes and hydraulic acids. Perhaps in the future, doors will open in terms of degree courses focusing on mineralogy and geochemistry for the environmental, industrial and mining sectors.

The statistical and GIS methods presented have a wide application in environmental studies, although they were fundamentally related to prospective applications in this case. The presentation of new data is also noteworthy.

José Maria Filippini Alba

Pelotas, May 5, 2017

CHAPTER 1

From prospecting to the environment

In the 20th century, geochemical prospecting developed dramatically with the aim of locating and quantifying mineral deposits, mainly metallic (GOVETT, 1983). Current sediment geochemistry has been used in various prospecting programs, with significant results even in areas that are poorly known from a mineral point of view. However, in the 1980s, there were changes in the direction of research, either due to the advance of environmental problems or the fall in the value of metals (saturation of stockpiles, emergence of alternative materials...), giving rise to environmentally-oriented research (THORNTON, 1983). As a derivative of this process, Medical Geology emerged, a discipline that studies the relationship between geological materials and processes and the incidence and space-time distribution of diseases in animals and humans (UNICAMP, 2003).

As a result of this process, several countries began compiling, updating and reprocessing the available geochemical bases. For example, Great Britain has produced several geochemical atlases at different scales, involving stream sediment and water samples for major and trace elements (THORNTON, 1983). A summary of the main results:

The three major British geological-physiographic units have shown a characteristic geochemical response, with the most contrasting features corresponding to the unit covered by Devonian and Tertiary sediments, compared to the two where older crystalline rocks predominate. These units control the geochemical variance of water, soils and vegetation based on the lithotypes of the basal rocks; however, control can fail due to drainage conditions, Eh/pH values or other factors. Copper (Cu), Cobalt (Co), Iron (Fe), Manganese (Mn), Molybdenum (Mo) and Zinc (Zn) are essential elements for agriculture, but Co, Cu, Mo and Zn, along with As, Cadmium (Cd), Chromium (Cr), Nickel (Ni) and Lead (Pb) can become harmful if certain thresholds are exceeded. Cd and Pb are indicators of regions contaminated by industry and mining.

Darnley (1990) referred to the IGCP 259 project, "International Geochemical Mapping", pointing out that surface geochemistry is an essential component of any comprehensive description of the natural environment. Although geological, pedological and geophysical maps of the Earth have been drawn up at various scales, there is no global geochemical information. Darnley et al. (1995) made methodological and technical recommendations for the construction of a global geochemical database.

Sampling methods were discussed, taking into account the different climates and landscapes of our planet . For temperate climates, surface waters, stream sediments and soils were recommended as collection materials. *Stream* sediments were categorized according to three sampling modalities:

floodplain sediments, *overbank sediments* and *stream sediments*. *Floodplain sediments* are alluvial sediments accumulated on the banks of large rivers, representative of river basins over 1000 km^2 and relevant on a continental or global scale. On the other hand, *overbank sediments* are sampled in low-order watercourses, which correspond to small basins (<100 $km^{2)}$ and are suitable for regional prospecting. *Stream sediments* are sampled in the beds of low-order streams. Other interesting considerations are: (1) 33% of the planet's continental surface has been surveyed by airborne gamma spectrometry (K, Th and U) and only 19% by conventional geochemistry; (2) when considering the analytical techniques, the materials sampled and the chemical elements analyzed, the situation becomes even more critical in terms of integration. Only two elements have always been considered: Cu and Zn.

Davenport (1993) mentions that regional geochemical surveys in Germany, Canada, China, Scandinavia, Greenland and Wales covered territories ranging from 1000 km^2 to 5 million $km^{(2)}$. In general, sediments were sampled from streams, which were compared with samples from watercourses, lakes or tillites, depending on the characteristics and needs of the countries involved. The number of elements analyzed varied between 20 and >30, with sampling densities ranging from one sample per km^2 to one sample every 30 km^2, depending on the size of the territory assessed. Geological aspects and mineral occurrences controlled the geochemical variance, but environmental problems and diseases were also highlighted by some geochemical associations.

In recent years, trace element geochemistry has evolved to improve data modeling methods (LANCIANESE & DINELLI, 2015), to geochemistry of specific elements targeting the environment and health (FIGUEIREDO et al., 2012; BOWELL et al., 2014), making it possible to carry out studies with spatial resolution below nano-particles considering diverse materials and isotopes (REICH et al., 2016). At the renamed "Goldschimidt" event, named after one of the forerunners of Geochemistry, subjects such as biogeochemistry, modeling of the atmosphere and oceans, studies of the crust and the solar system, continue to captivate the world's academic audience.

Thus, past geochemical prospecting data can be updated considering low-density sampling and involving a large number of trace elements, or can be reprocessed with new forms of modeling and related to environmental and human health problems. Data of this type has been collected in Uruguay, mainly in crystalline terrain (Precambrian), with approximately 25,000 samples, covering a territory of equal size in km^2, with 22 elements being analyzed by direct current plasma (DCP) spectrometry; information that was not recorded by Darnley et al. (1995), when they compiled the condition of geochemical prospecting projects in the world as a whole. Some of this data was considered by the author in his doctoral thesis (FILIPPINI ALBA, 1998) and is commented on here, along with various aspects of applied geochemistry currently in force. Applied geochemistry means studies involving surface geochemistry, geochemical prospecting and environmental geochemistry.

Several organizations around the world have established thresholds or tolerance values for potentially harmful elements in water, sediments and soils (National Environmental Council - Brazil, *Geological Survey of Finland, Ontario Minister of Environment* - Canada, *US Environmental Protection Agency, etc.)*, so geochemical data of prospective origin can be useful as a form of reference.

Medical Geology

The definition of Medical Geology was mentioned at the beginning of the chapter, according to UNICAMP (2003). These authors added: "A large body of evidence indicates significant effects on human health as a consequence of man's interactions with nature, with records dating back close to 10,000 years. Although the links between the physical environment and human disease have been recognized for a long time, the momentum built in recent years is moving towards solidifying and formalizing the study of these interactions. New techniques such as remote sensing and atomic spectrometry allow the world's scientific community to thoroughly analyze aspects related to environmental health."

Thornton (1983) indicated four well-recorded cases of the causal relationship between human health and the abundance of trace elements in the surrounding environment, i.e. in soils, plants, bodies of water and the atmosphere: (1) Populations with soils deficient in Iodine (I) show elevated thyroid growth statistics; a problem that can be minimized by adding NaI to table salt. (2) Regions with adequate levels of Fluorine (F) in drinking water show a 50% reduction in the rate of dental decay, while excess F leads to dental problems (fluorosis). (3) Soils depleted in Selenium (Se) have caused the development of a peculiar myocarditis in Chinese boys. (4) Arsenic (As) toxicity or skin cancer have affected Taiwanese who drank water from contaminated wells.

LÂG (1991) confirmed similar cases for F and I in regions of Scandinavia, for example, with the occurrence of fluorosis in Iceland related to volcanic emanations. Se levels below 0.025 $mg.kg^{-1}$ in grains in an extensive strip that crosses China in a southwest - northeast direction, overlapped with areas of occurrence of Keshan disease (cardiomyopathy) and Kaschin-Beck (osteoarthropathy) as presented by Tan et al. (1988 *apud* DARNLEY et al., 1995). Ng et al. (2003) mentioned a global health problem caused by As from natural sources. Occurrences have been recorded in 14 countries located in North and South America, Asia and Europe, involving millions of people affected by the occurrence of groundwater and coal deposits contaminated with this element.

In agronomy, biology, geochemistry and geosciences in general, it is common to use the terms macro- and micro-nutrients, heavy metal and essential or trace element. The former are differentiated by a scale factor, since living beings need a few milligrams of micro-nutrients a day, but more than 100 milligrams a day of macro-nutrients. "Heavy metals" are chemical elements with an atomic weight between 63 and 200, a density greater than 4 $g.cm^{-3}$, involving a metallic (Cd, Cr, Cu, Hg, Pb, Zn...), metalloid (As, Sb, Te...) or non-metallic (Se) character, whose availability is beneficial in small quantities for living organisms, but become toxic when certain thresholds are exceeded. The term "trace" can be confusing when considered with reference to the earth's crust or the biosphere; in both situations it means that it is found in low quantities, below 0.1%. However, confusion arises for elements such as Si or Fe, whose abundance in the crust varies between 27 to 32% and 4 to 7% respectively (GOVETT, 1983), but are micronutrients in the biosphere.

Adriano (1986) defined the concept of essential element for higher plants, considering three conditions: (i) The absence of the element causes abnormal growth, failure to complete the life cycle or premature senility and death. (ii) The effect must be specific and the element irreplaceable. (iii) The effect must be direct on some aspect of growth or metabolism. Similar concepts are considered for animal nutrition. The author describes the essential-beneficial or toxic nature of 30 trace elements. As, Co, Cr, Cu and F have both conditions, a fact explained by the concentration or the chemical species involved; for example, Cr^{+6} is highly toxic, but Cr^{+3} has low toxicity. Mn is phytotoxic in acidic soils, i.e. with a pH below 5. As, Be, Cd, Co, Cr, Ni, Ti and V are indicated as possible carcinogens, while Au, Br and Li have already been used in therapeutic treatments.

Shacklette et al. (1970) analyzed the influence of geochemistry on the mortality rate from heart disease in the state of Georgia, USA. Soil, vegetable and tree samples were collected from counties with high and low mortality rates from heart disease. It was concluded that trees would be more sensitive than vegetables for studies of this type; moreover, the differences in mortality rates would be due to deficiencies of essential elements and not due to the occurrence of toxic elements. Bowie and Thornton (1986) reported a negative correlation between the hardness of drinking water and mortality rates from heart disease in a study of 253 towns in Great Britain, which produced contradictory results on a local scale. The five factors that would explain the geographical variations in this index would be: water hardness, rainfall, temperature, percentage of manual workers and having one's own car.

Additional concepts on the subject can be found at: www.mineropar.pr.gov.br/modules/conteudo/conteudo.php?conteudo=129www.cprm.gov.br/publique/gestao-territorial/geologia-medica-41www.medicalgeology.org

CHAPTER 2

Statistics and geospatial analysis in Geochemistry

Statistics has been widely used in geochemical prospecting due to the accumulation of multi-element data, whether due to variations in the sample matrix, analytical methods or the nature of the projects involved. One of the main applications includes univariate procedures for differentiating anomalies in relation to background content or geochemical *background*, through the use of frequency histograms and conventional statistics (HOWARTH, 1983). In general, the background content can be considered as the average or median of a geochemical population. In practice, there may be groups of samples with different geochemical responses to a given variable, so that when they are superimposed on a frequency histogram, multimodal behavior occurs. Multi-element associations represent another way of processing the *background from* a multivariate perspective, which is very useful, especially when the association can be interpreted from an environmental, geological or production perspective.

Several authors agree on the need to evaluate data geospatially, considering not only regional averages to represent the *background*, but also local effects (CHORK & GOVETT, 1979; STANLEY & SINCLAIR, 1987; ROQUIN & ZEEGERS, 1987; CHENG et al., 1996). Robust statistics has also been applied (ZHOU, 1985; GARRETT, 1989), as it uses methods to identify and eliminate *outliers*, i.e. samples that are distant from or inconsistent with the general behavior of the population.

Geographic information systems (*GIS*[1]) make it possible to manage and process geospatial information, with a strong application for environmental sciences and geochemical prospecting. They are complex computer systems that process several layers of information simultaneously, as if they were transparent maps visualized by superimposition. However, the layers of information can be processed individually and integrated, resulting in new layers of information according to the user's criteria. Its main applications include geo-referencing, interpolation methods[2] and *multi-criteria* decision making (BONHAM-CARTER, 1994), also known as *GIS-MCDA* (MALCZENWSKI, 2006).

Subpopulation, anomaly and *outlier*

A group of samples with characteristic and homogeneous geochemical behavior, for example, with differentiated contents in relation to the rest of the population for certain elements, can be defined as a subpopulation. Depending on the spatial distribution of the samples and the area covered, this

1 We prefer to use the English abbreviation *GIS* for *Geographical Information System*, which is better known in academic circles.
2 A procedure that transforms point information into a continuous surface.

subpopulation could characterize a regional anomaly, a local anomaly or another type of phenomenon (including sampling errors, surface processes, contamination, etc.). It is usually interpreted that regional anomalies are spatially related to geological formations and local anomalies to the occurrence of specific mineralizations or lithologies (lithological anomaly). Typical examples of anomalies related to metallic sulphides are Cd - Zn (blende), Ag - Pb (galena), and in the case of lithotypes, Ba - Pb in granites and granitoids and Fe - Co - Cr - Ni in mafic rocks.

In geochemical prospecting, samples that are enriched in some element in relation to their surroundings (positive anomalies) are generally considered to be anomalies. However, there are situations of impoverishment (negative anomalies). In the latter case, an example in the environmental context is the low levels of selenium in soils in China, as mentioned above (DARNLEY, 1995).

In the context of statistics, *outliers* are defined as samples with behavior that is distant from or inconsistent with the total population (AFIFI & AZEN, 1972). It should be considered that every geochemical anomaly (positive or negative) is an *outlier*, however, the reciprocal is not true. For example, a succession of analytical or methodological errors can lead to the generation of *outliers*. Jolliffe (1986) suggests using scatter diagrams to discriminate *outliers*. In Figure 1, the Pearson correlation coefficient for Cr - Ni is 0.981. However, considering the samples inside the ellipse as the standard behavior, three *outliers* of different characteristics can be discriminated (O_1, O_2 and O_3). o_i is positioned close to the standard subpopulation *(background)* and deforms the linearity pattern. These *outliers* are difficult to discriminate by conventional univariate methods (histograms) and are called scale and orientation *outliers*. O_2 and O_3 are quite far from the *background* (ellipse), but are aligned with it, i.e. without reducing the degree of correlation, and are therefore called scale *outliers*. By eliminating o_2 and o_3, the Pearson correlation coefficient drops to 0.936. By also eliminating o_1, i.e. considering only the *background* subpopulation, the correlation coefficient increases slightly to 0.941. When the *background* population is non-linear, the *outliers* should be analyzed carefully, because depending on their degree of distance, linearity can be induced, resulting in a false correlation (when eliminating the *outliers*, the correlation coefficient drops sharply).

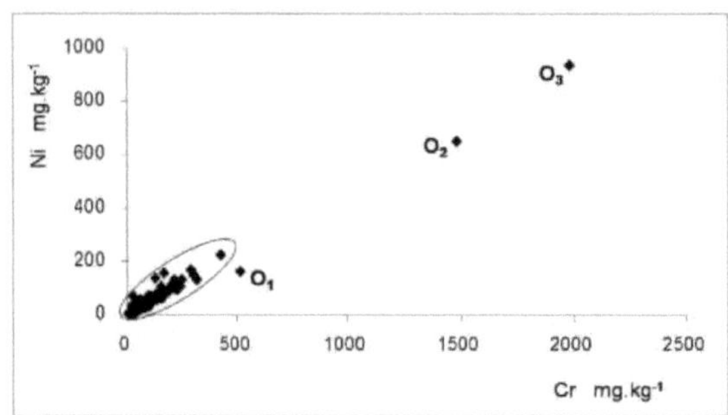

Figura 1. Cr-Ni dispersion diagram for 1935 terrace sediment samples in the Uruguayan crystalline.

The introduction of geochemical ratios in dispersion diagrams makes it possible to highlight the variability of populations, a procedure often used in Ichthyochemical studies (ROLLINSON, 1993). Figure 2 shows three subpopulations (ellipses A, B and C). Some samples retain the transitional behavior (intersections between ellipses), but even so, the *outliers* OA and oc are discriminated.

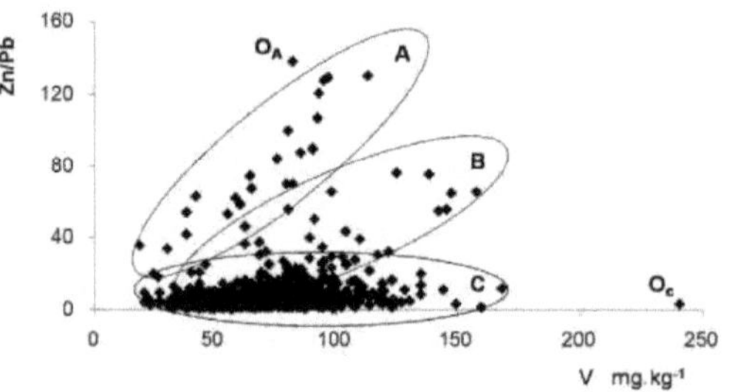

Figura 2. V-Zn/Pb dispersion diagram for 1935 terrace sediment samples in the Uruguayan crystalline.

When generalizing the above concepts to the multivariate case, we can expect an increase in the complexity of the situation described, with an increase in the number of subpopulations and *outliers*. The latter are detrimental to the application of multivariate procedures and should be eliminated. In addition to the use of graphical methods, there are multivariate procedures for identifying *outliers*, such as the Mahalanobis distance (AFIFI; AZEN, 1972; GARRETT, 1989) or principal components (JOLLIFFE, 1986). If there are several subpopulations, they must be identified and processed independently.

Statistical methods

Statistical methods are classified in several ways: (1) By the number of variables considered (univariate or multivariate). (2) The use or not of parametrization. (3) The elimination or not of *outliers* (mentioned above), known as "robust statistics". Non-parametric" statistics transform the variables into ranks, so it is not necessary to adjust them to a specific distribution law (Gamma, Normal, Student, etc.). Davis (1986) describes various statistical methods in the geological context, as does Howarth (1983) for geochemical prospecting.

In environmental applications related to impacts by potentially harmful elements and in geochemical prospecting, univariate statistics makes it possible to establish the preliminary parameters of populations (extremes, averages, quartiles, variance...), as well as to analyze the fit to distribution laws by means of histograms and other graphs.

For current sediment geochemistry, it is common to adjust to the Log-normal law. In this sense, the use of transformations, using the mean and standard deviation, yogarithms, square root or other types can help to reduce variability, so that the behavior of the *background* population is "adjusted".

Analysis of variance (ANOVA) makes it possible to study variability related to analytical and methodological errors, the influence of sample type and geology and other factors depending on the experimental design applied. Examples will be presented in the following sections.

In parallel with the study of the influence of geology, *dot* maps can be analyzed, based on thresholds calculated by geochemical variance or considering intervals according to subpopulations, especially in the case of multi-modal histograms. The term *natural breaks* is used by some *software* to define this last modality. It should be noted that these maps represent the preliminary phase of geospatial information analysis. In this context, it is possible to use interpolated maps, i.e. where point information is transformed into an image, a procedure that can be carried out using a GIS or geostatistical applications.

The analysis of dependencies between variables must take into account correlation coefficients to quantify the degree of correlation, without forgetting to look at scatter diagrams to eliminate *outliers*, as mentioned above. If parametric methods are used, the effect of *outliers* is reduced. So far, the methods have been mentioned sequentially, according to a specific geochemical study. The next steps concern multivariate statistics.

Although regression models can be run for a binary relationship, establishing multivariate models to explain a variable is more efficient. To do this, you can use Principal Component Analysis (PCA) or Factor Analysis in R mode, which allow you to establish multivariate associations.

Principal component analysis (PCA) is a mathematical method for extracting the eigenvalues and eigenvectors of the variance-covariance matrix or the correlation matrix. In practice, the N original variables are transformed into N new variables, which are the principal components. Each principal component is a linear combination of the original variables, preserving the total variance of the data.

The coefficients of the Iinear combinations (*loadings)* can be interpreted as the weight of the original variable in relation to the respective principal component.

In a similar way, but involving statistical criteria, factor analysis in R mode assumes that the total variance of a set of N variables is explained by a set of K variables (K< N), called factors, and by an additional component related to the variance not explained by the factors and related to sources of error. Thus, the principal components and factors can be considered "multivariate associations" of the original variables, making it possible to establish a direct relationship with geological phenomena. Jolliffe (1986) points out that the first principal components are related to global phenomena in the data.

Cluster analysis and discriminant analysis aim to characterize specific subpopulations of samples. They are complementary techniques, since the first classifies samples into groups and the second establishes discrimination between groups.

This section provides a brief explanation of statistical methods in applied geochemistry, considering the sequence of execution from the author's perspective.

Geospatial analysis

Due to the large number of ways of processing geochemical data through geospatial analysis, this section considers an example of decision making using Geographic Information Systems (*GIS)*. Initially, fuzzy logic was applied, which uses logical operators (AND, OR, NOT....), but also weights the variables continuously in the range 0 to 1. The method applied made it possible to locate and quantify the sectors with the greatest environmental risk of contamination by As and Pb, which are potentially harmful elements.

The history of mining in the Ribeira Valley dates back to the 1600s, with intense exploitation of Pb for practically the whole of the last century, and the occurrence of natural anomalies of As, Cu, Cr, Ni and Zn, all of which are potentially harmful to human health and the physical environment. This, together with regional economic and environmental aspects, represents an ideal framework for diagnostic research and risk assessment.

The study area covers the 1/50,000 sheets Iporanga and Gruta do Diabo, located in the Ribeira de Iguape river valley, on the border between the states of Paranà and São Paulo (Figure 3). Iporanga and Barra do Turvo are the main urban centers in the sector, connected by state highways.

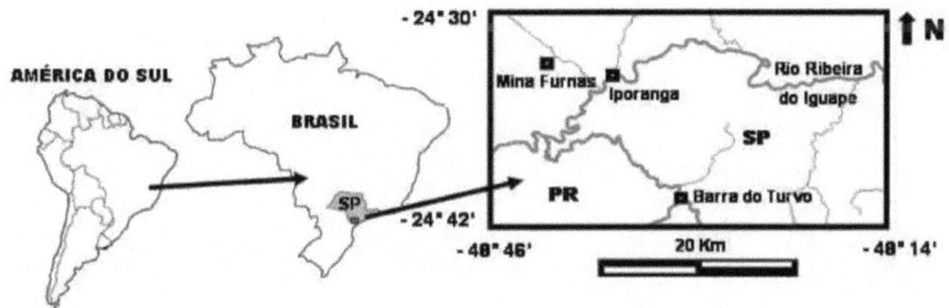

Figure 3 - Location of the study area in the Ribeira Valley.

Geochemical data from stream sediments collected by the Technological Research Institute of the State of São Paulo in 1985 in the Ribeira Valley, SP (sheets 1/50,000 Iporanga and Gruta do Diabo, unpublished study) were used. The fraction below 80 mesh was considered for several elements, but only As and Pb were used, analyzed by atomic absorption spectrometry and optical emission spectrometry respectively.

From the Landsat scene corresponding to orbit 220 and point 77 of the ETM+ sensor from September 26, 1999, the sectors showing exposed soils, i.e. absence of vegetation cover, were extracted. The slope data obtained from the SRTM digital elevation model, the As and Pb contents interpolated by the inverse square of the distance and the exposed soils were integrated using fuzzy logic in a GIS environment (ArcView®), in order to assess areas of potential risk to the environment and human health, a GIS-MCDA type process (MALCZEWSKI, 2006).

The geochemical data was compared to the US Environmental Protection Agency (USEPA) and Ontario Ministry of the Environment (OME) continental sediment guide levels. The criteria used to define the fuzzy variables are shown in Table 1 and the results of the integration in Figure 4.

The model adopted considered that the areas at risk were those with exposed soil (in red in the top left figure), but with increased mobility when there was a greater slope, due to the greater risk of erosion. The higher the As or Pb content, the greater the risk of dispersion, hence the gradual increase in the weights (Value 3 and Value 4). The sectors with the highest risk are indicated with blue rectangles.

Table 1. Criteria adopted to classify the data in the fuzzy procedure. CR = Class represented.

Soil map		**Slope map**		**Map of As**		**Map of Pb**	
CR1	**Value1**	**CR2**	**Valor2**	**CR3#**	**Value3**	**CR4#**	**Value4**
Vegetation	0	0 - 4°	0	0 - 10	0	0 -45	0

Soil	1	4 - 9°	0,1	10 - 20	0,2	45 - 90	0,1
		9 - 13°	0,2	20 - 35	0,4	90 - 130	0,2
		13 - 17°	0,4	35 - 70	0,7	130 - 220	0,6
		17 - 22°	0,5	> 70 ppm	1	> 220	1
		22 - 26°	0,6				
		26 - 30°	0,7				
		30 -35°	0,8				
		35 - 39°	0,9				
		39 - 43°	1				

mg.kg^{-1}

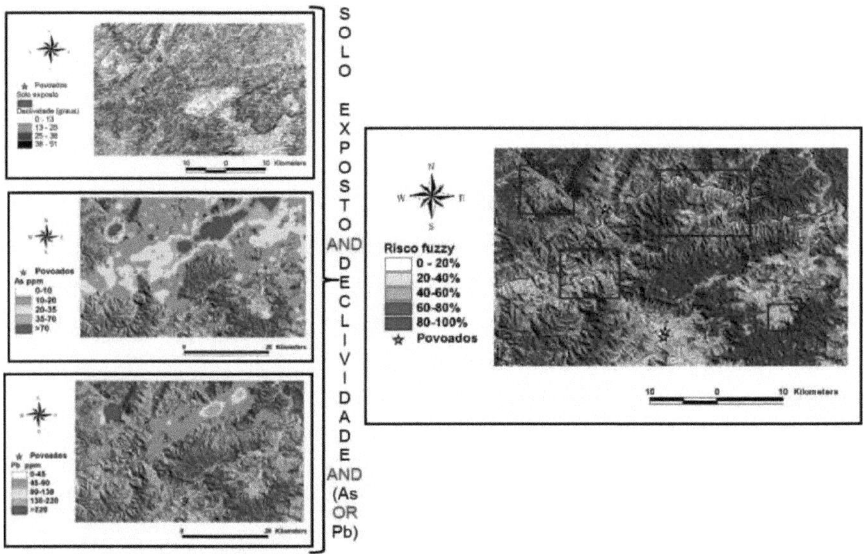

Figure 4: Representative diagram of the GIS-MCDA process using fuzzy logic in the Riveira valley region, SP, Brazil. Source: FILIPPINI ALBA & SOUZA FILHO, 2010.

Integration according to Boolean logic, also carried out in a GIS environment, makes it possible to illustrate and compare both methods (Figure 5). It can be seen that the information plans are binary, i.e. Yes (value 1) and No (value 0), as is the result of the integration. This condition can be modified by considering different options for the binary combinations in the integration result, for example: 0000, 0001, 0010, ...0011, 0101.... 1111 (BONHAM-CARTER, 1994). As the product was used as the form of integration (intersection), the presence of the value 0 for a single plane of information eliminates the other themes, so the result is also binary. But if the sum were used as the integration operator, the final result would be 0, 1, 2, 3 or 4, so it would be possible to consider intermediate stages with the result expressed discretely but not binary.

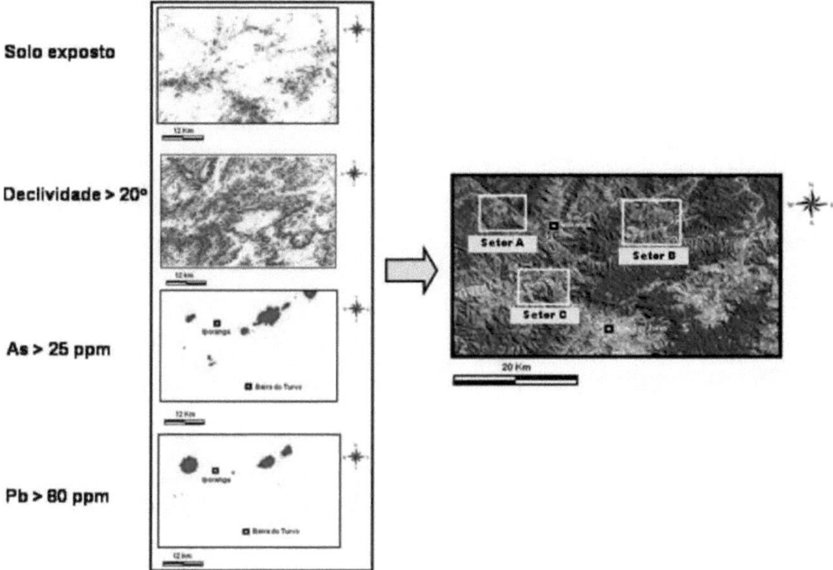

Figura 5. Representative diagram of the GIS-MCDA process using Boolean logic in the Riveira valley region, SP, Brazil. Integration by multiplication.

In both cases, two of the mapped sectors coincided with positive results from previous toxicological studies (those located to the north). In addition, for the Boolean procedure, one of the sectors was not considered due to the fact that As was in the 20 - 25 $mg.kg^{-1}$ range and Pb had a content of less than 45 $mg.kg^{-1}$ (Figure 4).

CHAPTER 3

Characteristics of the study areas

This chapter presents two of the areas sampled by the regional geochemical survey of Uruguay (1980-1988). By way of introduction and given its importance from the point of view of interpreting geochemical data, a brief description of the geology of Uruguay is also presented, being detailed, as far as possible, in the aforementioned areas. Known mineral occurrences will be reported considering their history of discovery and mining.

Location of the study areas

The next chapters will present geochemical studies in Uruguay, considering two study areas (Figure 6):

(1) *Maria Albina - Piraraja - José P. Varela - Zapicân* (MPVZ) covers four 1:50,000 sheets (*Gutiérrez, José P. Varela*, *Piraraja* and *Zapicân*) totaling approximately 2800 km^2, where 2172 samples of current sediments were collected in an area covering 2200 km^2. Filippini Alba (1998) was used as a reference. The José P. Varela sheet is located on the border of the departments[3] of *Treinta y Tres* to the north and *Lavalleja* to the south, with the remaining sheets falling completely within the latter.

(2) The area is located on the outskirts of the town of *Isla Patrulla*, in the department of *Treinta y Tres*, and is roughly one third of the 1:50,000 sheet of the same name (National Mapping Plan of the Military Geographic Service, Uruguay), where 220 quartz veins and 341 samples of current sediments were sampled over an area of approximately 260 km^2. The information compiled and processed by Filippini Alba (1992) was taken into account.

The digital media (2014) indicates that the department of *Lavalleja* has an extension of 10,016 km^2 with 60,925 inhabitants and in the case of *Treinta y Tres*, 9,676 km^2 and 49,318 inhabitants respectively. The departmental capitals are Minas and Treinta y Tres, the main regional urban centers, both of which are located outside the study areas.

Total gamma radiometry with a portable scintillometer and pH (universal indicator paper) were also determined at each current sediment sampling point.

For both the stream sediments and the quartz strands, 22 elements were analyzed [4] by direct current

3 The territory of Uruguay is divided into 19 "departments", whose area varies between 530 km2 (Montevideo) and 15,438 km2 (Tacuarembó).
4 Iron (Fe), Manganese (Mn), Phosphorus (P), Antimony (Sb), Arsenic (As), Boron (B), Barium (Ba), Beryllium (Be), Cadmium (Cd), Cobalt (Co), Chromium (Cr), Copper (Cu), Molybdenum (Mo), Niobium (Nb), Nickel (Ni), Silver (Ag), Lead (Pb), Tin (Sn), Vanadium (V), Tungsten (W), Yttrium (Y) and Zinc (Zn)

plasma emission spectrometry, also known as DCP (*Direct* Current Plasma) spectrometry, considering the fraction below 80 mesh. The samples were subjected to total digestion by means of a two-stage acid attack: (1) perchloric acid ($HClO_4$) at 140°C; (2) addition of hydrochloric acid and fluoride (HCl-HF) at 80°C.

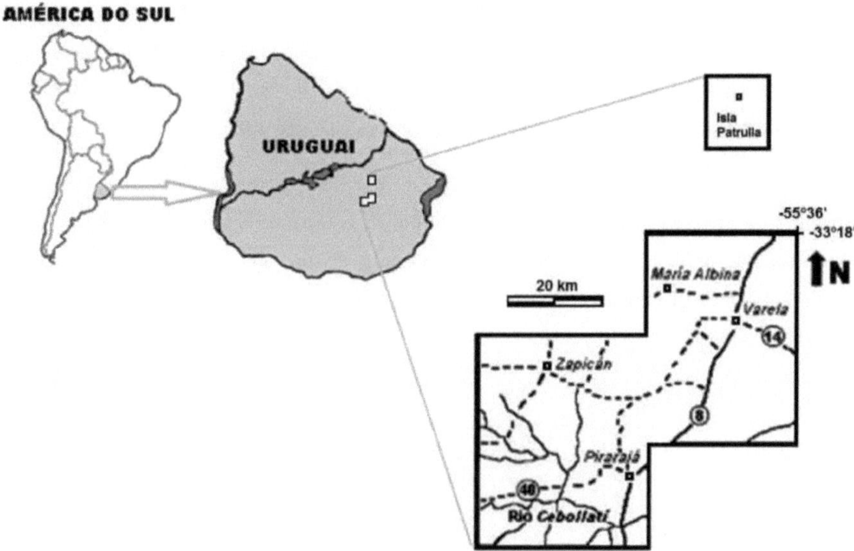

Figura 6. Location of the study areas considered in Chapters 4 and 5: *Isla Patrulla* and *Maria Albina - Pirarajà - Varela - Zapicân* (MPVZ), Uruguay.

Brief Geological Description

The geological map of Uruguay at a scale of 1:500,000 (PRECIOZZI et al., 1985, 1988) was conceived considering two orogenic cycles: an older cycle represented by the *Rio de la Plata* craton, and the Brasiliano cycle related to the Dom Feliciano Belt (FRAGOSO CESAR et al., 1987). However, in the transition zone of the rocks corresponding to each cycle, there are uncertainties, which have generated some controversies, expressed in unpublished reports: *Isla Patrulla* Formation, Amphibolite-Leptite Formation and *Pavas* Formation (Fesefeldt et al., 1988).

Bossi (1987) indicates the existence of a primitive core of more than 2 Ga, located in the transition region of the known orogenic cycles. Gómez-Rifas (1985) considers this region to be part of the *Rio de la Plata* craton, with Brazilian cover and deformation.

Masquelin et al. (2017) describe the Dom Feliciano Belt as a vigorous set of crustal blocks of "Paleoproterozoic - Archean" age subjected to orogenic processes of approximately 2.2-2.0; 1.7 and 0.6 Ga. Each "Paleoproterozoic - Archean" block is covered by successions of Meso to

Neoproterozoic schist belts. However, the entire structure of the Belt is linked to the last phase of the Ediacaran transpressive collision and the respective unearthing process, from 0.55 to 0.58 Ga, developed after a prolonged episode of multiple orogenic-type subduction. On the other hand, there are tectonic structures that suggest the occurrence of a *fold-and-thrust* belt[5] in the origins.

The MPVZ region is located in the aforementioned transition zone, and the combined geology of the four sheets was compiled by Fesefeldt et al. (1988), taking into account unpublished geological maps (Figure 7). The undifferentiated sedimentary cover includes several Cenozoic formations that occur near the *Cebollati* River and to the east of the study area. The vacuolar Jurassic basalts of the *Puerto Gómez* Formation occur in the vicinity of the town of *Pirarajà*.

Granites occur with different mineralogical compositions and forms of *emplacement*. The *Pirarajà* granite is a batholith with an equigranular to porphyritic texture and an intermediate granite-granodiorite composition. To the south of the former, there are small bodies of microgranular leuco-granites with scarce mica.

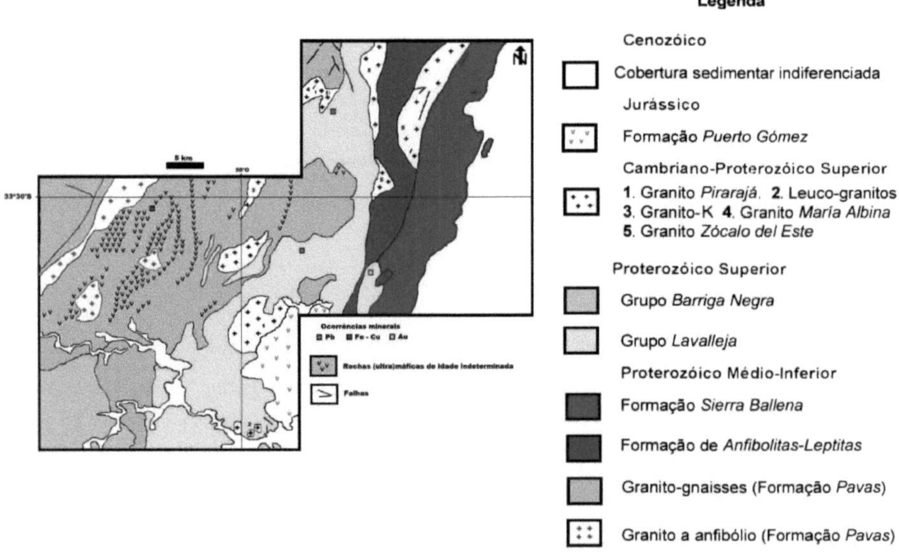

Figure 7 - Geological map of the MPZ region, Uruguay (Adapted from Fesefeldt et al., 1988).

To the north of the *Pirarajà* granite there are two granitic bodies with a granular texture, abundant microcline ($KAlSi_3O_8$) and little biotite. The *Maria Albina* granite is a rounded body with schistosity

---

5 Characterized by the lithostratigraphic content typical of marginal oceanic zones, the structure with low-angle reverse faults and folds with a notorious vergence towards the neighbouring continent/craton (*www.neotectonica.ufpr.br/aula-geotectonica/aula6.pdf*).

on the edges, containing perthitic microcline with small plagioclase crystals, quartz with two different textures and abundant biotite. The three bodies corresponding to the *Zócalo del Este* granite, of sub-alkaline, granular and microgranular composition, show intense schistosity at the edges, deriving into granite-gneisses, where elongated biotite bands alternate with quartz-feldspathic levels. Mafic veins with a porphyritic texture and phenocrysts of amphibole and olivine, with a width of less than 20 m and a preferential direction of N30°L, occur embedded in the rocks of both orogenic cycles.

The *Barriga Negra* Group is discordant with the *Lavalleja* Group and is made up of sedimentary rocks with an almost horizontal slope, including limestone, siltstones, sandstones, conglomerates and breccias. The latter are of two types, some with quartzite clasts similar to those of the *Lavalleja* Group, the others with granite, volcanic and metasediment clasts, thus characterizing two sedimentation palaeoenvironments, one calm and the other agitated.

The *Lavalleja* Group in Uruguay is characterized by a band 5 to 20 km wide heading N20°L, which rises near the *Rio de la Plata* and disappears into the Gondwana cover to the north, outcropping again in southern Brazil. In the region studied, limestone and calcofilites predominate, with intercalations of sericitic-chloritic phyllites, quartizites, sandstones and mafic volcanic rocks.

The *Sierra Ballena* Formation is a sinistral shear zone of continental magnitude that controls the structural geology of the Uruguayan-South-Riograndense Scheldt and its Phanerozoic cover, determining the distribution, nature and age of the adjacent lithologies (GÓMMES-RIFAS, 1995). The occurrence of hot millonites (550 °C) resumed in shallow levels of the crust indicates a broad spectrum of temporal evolution, involving the collision period to late and post-tectonic periods.

Fesefeldt et al. (1988) defined the Amphibolite-Leptite Formation as the base of the *Lavalleja* Group due to its more intense metamorphism and close relationship. The amphibolites are microgranular, occasionally schistose, containing hornblende, plagioclase, epidote, actinolite and quartz. Leptites, on the other hand, are fine-grained and composed of feldspar and quartz. Gneisses also occur. The authors suggest an aulacogem/rift facies to explain the origin of the formation.

The Pavas Formation is characterized by alternating bands of mafic rocks and felsic rocks of kilometric width, N0°L - N30°L direction and medium-high grade metamorphism. The mafic rocks are amphibolites and amphibole schists, while the felsic terms are leucocratic gneisses and leptinites. An elongated granitic body in a N20°L direction and two more rounded bodies, composed of plagioclase, microcline, quartz, biotite and amphibole, located to the west and south of *Zapicàn*, have been mapped as Granite to Amphibole.

The geology around the town of *Isla Patrulla* is very similar to that described above (FESEFELDT et al., 1988). However, there is the *Isla Patrulla* Formation, a 1-2 km wide strip of amphibolites and quartzites, which extends to the north and south, possibly lying discordantly on top of the *Pavas* Formation (FESEFELDT et al., 1988). There is deformation in the rocks, where two types of amphibolite occur, one derived from ultramafic protolith rich in Mg and the other a product of metamorphism of basaltic or andesitic rocks. In addition, the *Fraile Muerto* fault has millonites similar to the *Sierra Ballena* Formation.

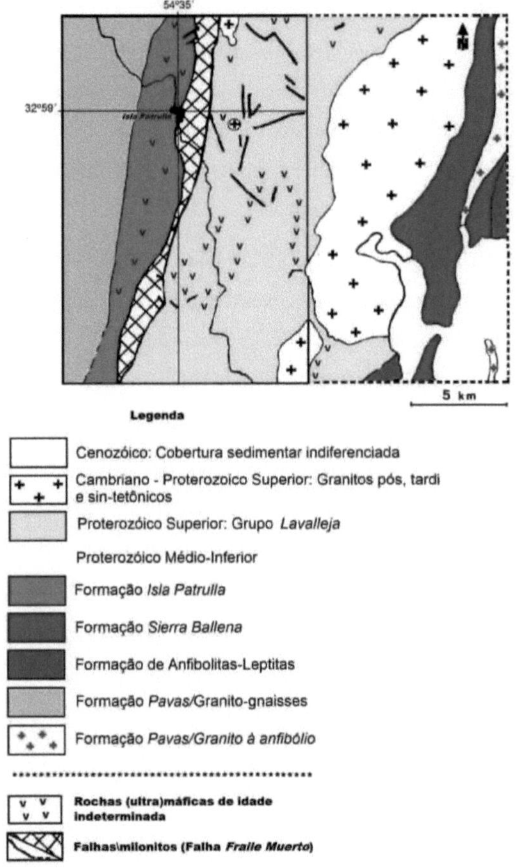

Figura 8. Geological map of the *Isla Patrulla* region, Uruguay (Adapted from Fesefeldt et al, 1988).

Mineral occurrences

There are four mineral occurrences in the MPVZ region, one related to the *Pavas* Formation and the other three to the *Lavalleja* Group (Figure 7). The Fe - Cu occurrence is located near the town of *Zapicân*, in the form of a quartz vein mineralized with pyrite - chalcopyrite and embedded in felsic gneisses. The galena (Pb) occurrences were prospected in the 1980s. In both cases, the bedrock is limestone and the mineralization is associated with silicified and ferruginous levels.

In the southwestern portion of the *Gutiérrez* sheet, there is an old gold mine (Au), which was exploited in the 1930s-40s. It is a quartz vein, violet in color and less than a meter wide, with an average Au content of 7 $mg.kg^{-1}$, with pyrite and chalcopyrite as accessory minerals. There are records of precious metal extraction in the order of 200 kg. In an unprecedented study of three soil transects, a positive anomaly was detected in the C horizon, with an *anomaly/background* ratio of over 1.5 for Au, Ba, Co, Mn, P and Pb.

The discovery of a ten-gram gold nugget in the schoolyard of the village of *Isla Patrulla* in the 1960s brought a "golden age" to the small town. An unprecedented study carried out in the 1980s, through alluvial prospecting, in approximately 90 watercourses in the southern part of the area considered here resulted in the occurrence of millimetric grains of Au in 60% of the surveys carried out. The positive occurrences were located on both sides of the *Fraile Muerto* fault.

The author of this book verified the occurrence of galena and pyrite in quartz veins and malachite in limestone in a geochemical anomaly in stream sediments located to the east of the town of *Isla Patrulla* in the 80s.

New unpublished reports at the end of the 1980s showed that the quartz veins with the highest probability of auriferous mineralization in the vicinity of the *Fraile Muerto* fault were associated with the presence of first-order shear faults with mafic embedments or the existence of these rocks in the vicinity or, when the embedments were more recent rocks, the occurrence of Cu and Pb minerals associated with pyrite and Fe oxides filling fractures and cavities.

CHAPTER 4

Geochemistry of current sediments

Methodological errors and influence of the material sampled

In geochemistry, it is common to insert replicates to evaluate methodological errors. Thus, replicates of the same sample, usually collected in larger quantities, are analyzed to evaluate the analytical error (by digestion and instrumental measurement) and samples collected over a short distance, every 50 to 100 meters, are used to evaluate the sampling error (how does the imprecision of the position affect?).

Filippini Alba (1998) indicated that Fe, Mn, P, Ba, Co, Cr, Cu, Ni, Pb, V, Y and Zn showed analytical errors of less than 10%, without their distribution being affected by the detection limits. Ag, As, B, Be, Cd, Mo, Nb, Sb, Sn and W, on the other hand, showed analytical errors of over 15%, with many samples below the detection limit. For this reason, the first group of elements would be suitable for continuing the geochemical study. However, P, Co, V and Y showed sampling errors of over 55% and Mn, Ba, Cr, Pb and Ni between 40 and 50%, which could be explained by abrupt changes in the lithologies, the nugget effect or errors in the collection method, for example, by considering cross-sectional rather than longitudinal sampling.

In regional geological surveys, materials from different origins are collected which can present different geochemical responses. For example, in the case of the crystalline terrains of Uruguay, the soils were depleted in relation to the sediments with regard to some chemical elements (Table 2). As the samples are spatially intermixed, the effect should be corrected if the data is worked on in an integrated manner.

Table 2. Medians of selected elements (concentration in $mg.kg^{-1}$) for different sampling matrices, considering geochemical prospecting data in the MPVZ region, Uruguay (Filippini Alba, 1998).

Material	NA	Mn	Ba	Cr	Pb	V	Zn
Bed sediment	34	799	531	30	13	58	54
Sediments on terraces	1933	893	515	32	13	63	57
Alluvial soils	176	722	437	25	11	53	47
Soils *in situ*	**29**	**550**	**390**	**19**	**10**	**47**	**38**

NA = number of samples.

One way of making this correction is to express the concentration as a ratio of the median of each group classified according to the type of sampling. Another way of correcting this is to eliminate the group with a differentiated signature, clearly the "Soils *in situ" in* this case. Throughout the text, in a generic way, due to the similar response, the other three groups will be identified as "stream sediments"; emphasizing that they are representative of small micro-basins with a sampling density of one sample per km^2.

Geochemical *background* and influence of geology

The median is a centered estimate, i.e. it is little affected by the extremes of the distribution, serving as an evaluation of the geochemical *background* a priori. The value of the extremes and the coefficient of variation allows us to predict the behavior of the distribution laws. For example, those variables with a coefficient of variation of 30-50% have normal or log-normal behavior (Fe, Ba, Cu...), while between 50-75% the distribution laws are deformed, with a log-normal fit in general (Mn, P, Cr, Ni...), showing tendencies towards various modes. For values above 75% of the coefficient of variation, there is a strong influence of the lower limits of detection, clearly observed in some situations (Ag, Cd, Sb and W), whose lower end and median are zero.

Table 3. Extremes, medians and coefficients of variation of major and trace elements in stream sediments in the MPVZ region, Uruguay (2172 samples).

Variable	Minimum	Maximum	Median	Coefficient of variation
Fe	0,8	8,4	2,7	34 %
Mn	74	15610	863	71%
P	99	2528	242	74%
Ag	0	1	0	500%
The	0	80	5	96%
B	0	19	8	43%
Ba	174	2272	506	35%
Be	0	6	1	43%
Cd	0	1	0	400%
Co	1	78	11	50%
Cr	8	1980	31	146%
Cu	2	60	18	33%
Mo	0	4	0	242%
Nb	0	13	0	243%
Ni	0	940	17	133%
Pb	0	116	13	50%
Sb	0	11	0	174%
Sn	0	77	3	118%
V	18	240	62	30%
W	0	13	0	281%
Y	8	266	20	54%
Zn	17	176	56	33%

Units (extremes and median): Fe in %, remainder in mg.kg^{-1}

The histograms confirm the above concepts (Figure 9). The Ni histogram is quite deformed in the normal case, but *there is* a better approximation with the log-transformation. A similar situation occurs for Pb, but with a strong deviation towards the central interval. The adjustment to lognormality is evident for Zn.

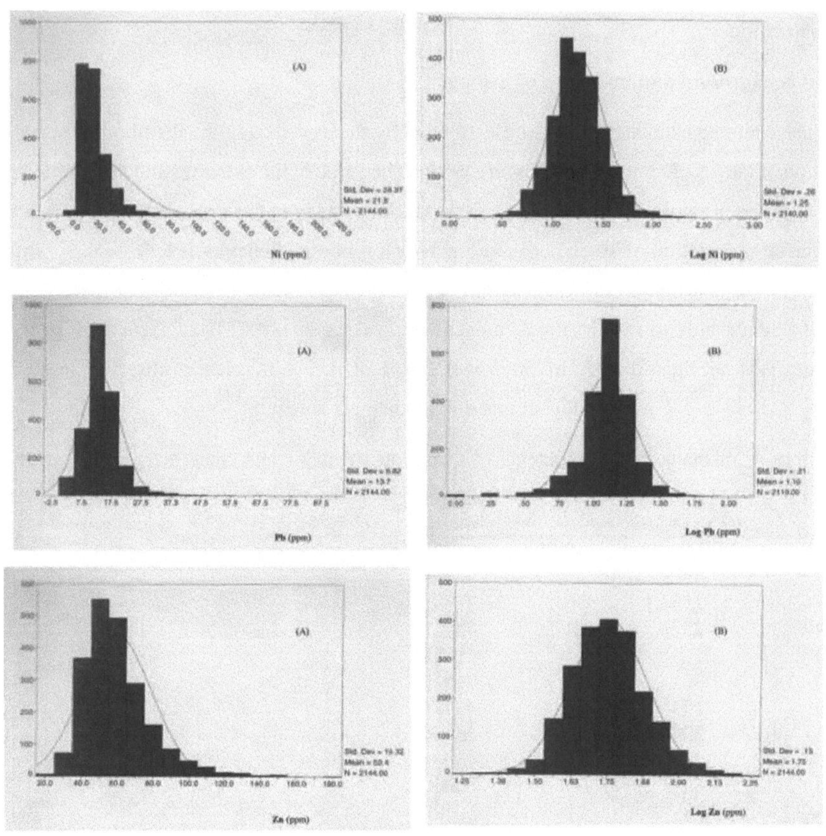

Figura 9. Histograms of Ni, Pb and Zn contents in stream sediments from the MPVZ region, Uruguay for 2144 samples (1% of *outliers* were discarded).

Geology affects geochemical variance, and there is a lot of evidence to this effect. Fe, Mn, P, Ba, Co, Cr, Cu, Ni, Pb, V, Y and Zn showed significance for the non-parametric Kruskall-Wallis statistical test with a significance of 1% (DAVIS, 1986), considering the comparison of means according to geological domains[6] . This means that at least one of the averages classified according to geology differs from the others. Geological domains can present high and contrasting medians, even in the context of the same formations, for example in the case of Fe, Mn, Ni, Pb and Y (Table 4). When expressing the values graphically, Fe and Mn behave similarly, as do Ni - Y and differently in the case of Pb due to the amphibolites of the Pavas formation.

Table 4: Selected medians for groups of stream sediment samples classified according to geological

6 Current sediment samples are classified according to the dominant geological domain.

domains with more than 85 individuals (N= number of samples).

Geological Domain	N	Fe (%)	Mn (mg.kg^{-1})	Ni (mg.kg^{-1})	Pb (mg.kg^{-1})	Y (mg.kg^{-1})
Sin, tardi and post granites tectonic[#]	151	2,2	700	17	15	19
Carbonàticas Grupo *Lavalleja*	385	3,4	893	16	13	19
Lavalleja Group shales	89	2,8	1103	18	14	18
Milonitos	180	2,3	893	11	14	17
Form. Amphibolites - Leptites	145	2,4	893	20	15	17
Amphibolites Form. *Pavas*	95	3,6	1051	37	8	38
Granite-Gneiss Form. *Pavas*	638	3,2	893	23	13	26

The *Maria Albina, Piraraja* and *Zócalo del Este* granites were integrated due to their similar geochemical response.

A similar situation applies to principal component analysis (PCA) (Table 5). For each lithotype, the *PCA* explained 68%, 61% and 71% of the total variance associated with syn-, late- and post-tectonic granites, volcano-sedimentary rocks (*Lavalleja* Groupand granite-gneisses/magmatic rocks (*Pavas* Formation), respectively (sum of VEX values). CP1 explains the largest percentage of the total variance. For granites and volcanic-sedimentary rocks, the result was similar, but for granite-gneisses/mafic rocks, there were changes with the participation of phosphorus, the absence of barium and the opposition of Pb. In any case, in all cases, the association expressed by CP1 seems to be associated with the occurrence of Fe-Mn oxides, probably derived from the mafic minerals incorporated into the parent rocks and this is undoubtedly a regional effect (*background).*

CP2 shows different associations in each case. For the Granites, it suggests a superficial phenomenon related to a second mineral phase containing Fe and Mn oxides that contrasts with secondary phosphates (?); in the other two domains, the contrast between the main rocks in each case prevails, limestone and schists in the *Lavalleja* group and felsic and mafic minerals for the *Pavas* Formation. A similar situation, but with different minerals, would explain CP3 in each case.

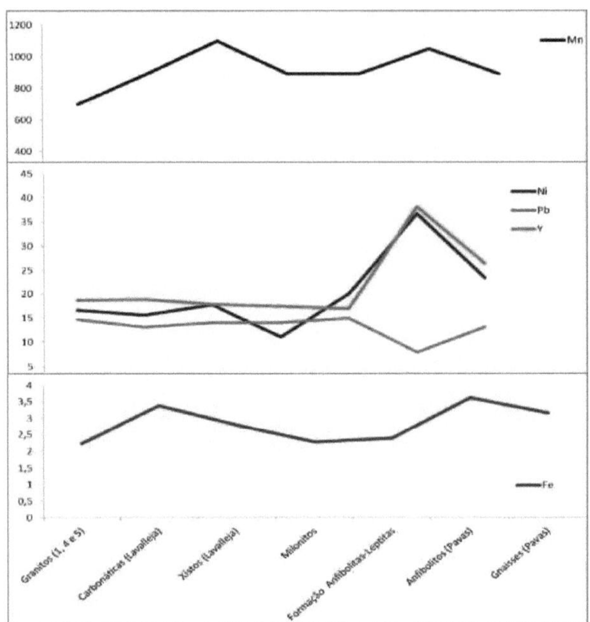

Figura 10. Median plots for groups of stream sediment samples classified according to geological domains with more than 85 individuals (N= number of samples).

Table 5 - Results of the PCA (R mode) considering multi-element geochemical data from current sediments from the MPVZ region, Uruguay, processed according to each geological domain.

Associated lithotype	**NA**	**CP**	**Variance Explained**	**Variables with *loadings* > 0.35**
Granite	172	1	46 %	Fe, Mn, Ba, Co, Cr, Cu, Ni, Pb, V, Y, Zn
		2	12 %	Fe, Mn, −P, Co
		3	10 %	Mn, P, −Pb, -Y
Volcano-sedimentary rocks	556	1	37 %	Fe, Mn, Ba, Co, Cr, Cu, Ni, Pb, V, Y, Zn
		2	13 %	Mn, P, −Cr, −V, Zn
		3	11 %	P, -Pb
Granite-gneiss/mafic	782	1	45 %	Fe, Mn, P, Co, Cr, Cu, Ni, −Pb, V, Y, Zn
		2	15 %	P, Ba, −Cr, −Ni,
		3	11 %	Mn, Co, Pb, -Y

NA = number of samples; CP = principal component. Note: The negative sign means negative charge, i.e. the opposite behavior of the element in relation to the PC and the positive elements (those without a sign).

As has been mentioned, statisticians consider principal component analysis or factor analysis to be variable reduction methods, since PCs and factors are new variables, expressed as linear combinations of the original variables, which can be interpreted as geochemical associations related to geological or environmental phenomena. When these methods are applied to the total population, the first PCs or factors are related to regional effects or the *background.*

Table 6 shows the *loadings* from Factor Analysis after VARIMAX rotation for the total population of stream sediments in the MPVZ region, Uruguay, considering 2128 samples, as 44 outliers were eliminated. To calculate the scores for each factor, the *loading of* each variable is multiplied by its value, and the products are added together. The VARIMAX rotation simplifies the interpretation of the original factors (more similar to the PCs) by "adjusting" the *loadings* to extreme values (0 or 1).

Table 6. *Loadings* for the factor analysis of the total stream sediment population with the elimination of 44 *outliers* (2128 samples) for the four-factor extraction after rotation.

Variable	Factor 1	Factor 2	Factor 3	Factor 4
Fe	**0,80**	**(0,33)**	**(0,28)**	-0,1
Mn	0,10	0,18	**0,93**	0,06
P	0,12	**0,69**	0,10	**-0,47**
Ba	**0,90**	**0,83**	0,16	0,2
Co	**0,56**	0,08	**0,75**	0,05
Cr	**0,90**	0,02	0,02	-0,15
Cu	**0,79**	0,12	0,15	0,07
Ni	**0,89**	0,05	0,18	-0,10
Pb	-0,08	0,01	0,09	**0,94**
V	**0,83**	0,15	0,19	
Y	**0,57**	**0,45**	-0,17	
Zn	**0,56**	**0,51**	0,23	
Variance explained	47%	13%	11%	7%
Interpretation	Traces adsorbed on Fe oxides	Surface phosphates	Adsorption on Mn oxides	P versus Pb contrast

Note: Bold represents significant elements, while brackets represent intermediate conditions.

Factor 1 represents an association of Fe, Cr, Co, Cu, Ni, V, Y and Zn which explains 47% of the total variance and can therefore be interpreted as the *background.* In a way, it is controlled by the occurrence of mafic rocks, such as amphibolites, which will result in the formation of surface Fe oxides through weathering. Factors 2 and 3 refer to surface phenomena, while factor 4 shows a strong contrast between two elements (P and Pb), whose levels are generally high for one of them, or at least in a representative portion of the area, and low for the other, or vice versa. When factor 1 was mapped, after interpolating the scores considering the inverse of the square of the distance with a radius of 6 km and an adjustment parameter of 1.2 km to define the weights, according to the Butterworth function, and assigning a color scale to the respective values, the continuous surface in Figure 11 was obtained.

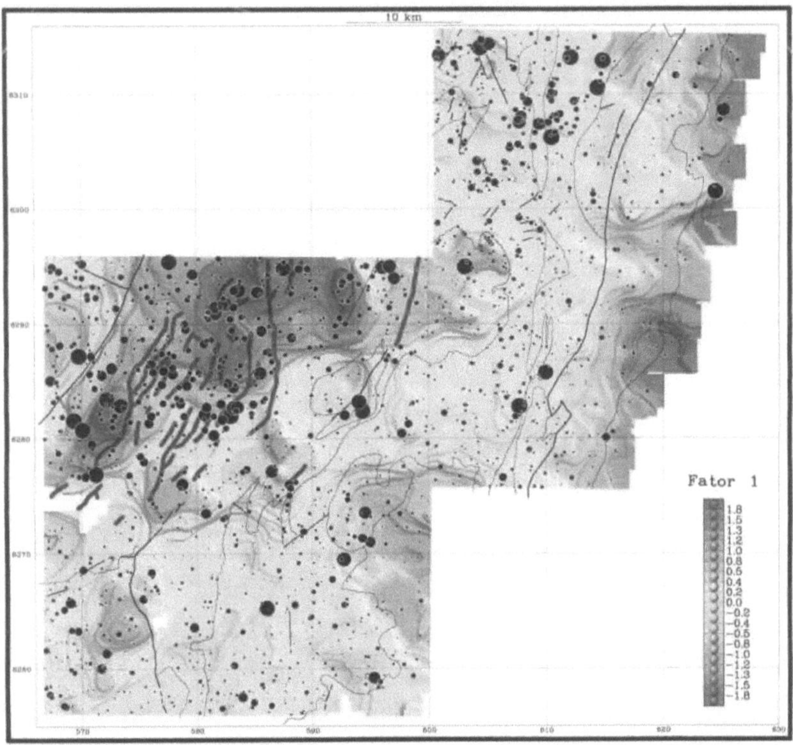

Figura 11. Map of Mahalanobis distance circles on the interpolated map of factor 1 (multivariate association interpreted as the geochemical *background*) by the inverse of the square of the distance in the MPVZ region, Uruguay. The fine and coarse polygonal traces represent the main geological features in the area (Figure 7). Factor 1 = 0.8 (Fe+Cu+V) + 0.9 (Ni+Cr)+ 0.6 (Co+Y+Zn). Map produced by the Geological Survey of Finland.

The highest values of factor 1 were associated with the amphibolites of the *Pavas* formation (lilac tone), *while* pink characterized the granite-gneisses of the *Pavas* formation and the *Barriga Negra* group. Yellow was related to the *Lavalleja* group, and in the Amphibolite-Leptite formation and the syn, late and post-tectonic granites it was mixed with slight greenish tones. Finally, in the *Sierra Ballena* (millonites) and *Puerto Gómez* (basalts) formations, greenish to light blue tones predominated, turning to intense blue (lower values) for the Cenozoic sedimentary cover. The map of points (Mahalanobis distance) will be explained and interpreted in the next section.

Another procedure applied to stream sediment data from the MPVZ region was integration with remote sensing data by means of unsupervised classification with 14 levels of information (FILIPPINI-ALBA, 1998; FILIPPINI-ALBA et al., 2001). Previously, 11 geochemical variables (total gamma radiometry and Ba, Co, Cu, Mn, Ni, Pb, P, V, Y and Zn contents) were interpolated by the inverse of the square of the distance with a pixel size of 200 meters and a radius of 1.5 km. Three variables were extracted from the orbital data of the Landsat satellite's TM sensor: band 4, the F factor and the H factor. The F and H factors are linear combinations of the thematic bands of the aforementioned sensor, evaluated by the *Feature-oriented Principal Component Selection* procedure (CRÓSTA & MOORE, 1989), which uses PCA, with the F factor relating to the content of Fe hydroxides and oxides and the H factor to minerals containing the OH molecule (clay minerals). Unsupervised classification groups together sets of *pixels* with similar characteristics, based on the statistics of each group and probabilistic criteria (CRÒSTA, 1993). Twelve classes were generated and grouped into four (Table 7), taking into account the spatial distribution of the original classes and their characteristics.

Table 7. Characteristics of the classes defined by digital processing (unsupervised classification).

Class	**Description**	**Area of Scope**
B	Low-contrast *background.*	34 %
E	*Background of* greater contrast, with contents enriched in Co, Cu, Ni, V and Zn compared to class B.	33 %
M	Two of the original classes are grouped together, with a low Ni content. One class showed enrichment of the F factor and Pb and the other of Co, Mn, Pb and gamma radiometry. Possibly related to mineralization.	5 %
L	It groups together seven of the original classes, depleted in Pb and enriched in Co, Cu, Mn, Ni, P, V, Ye Zn. Probably of lithological origin derived from the occurrence of mafic and ultramafic rocks (mainly amphibolites).	27%

When comparing the geological map (Figure 7) with the distribution of the four classes (Figure 12), it can be seen that class B is spatially related to the basalts of the Puerto Gómez Formation, the Sierra Ballena Formation and parts of the Lavalleja Group and the Amphibolite-Leptite Formation, while class E predominates mainly over the granite-gneisses of the Pavas Formation. However, the geochemical signature is not perfectly homogeneous, with both classes alternating within a given geological domain. The enrichment in Co, Cu, Ni, V and Zn of class E suggests an association with class L, which is due to proximity, because of the occurrence of mafic and ultramafic rocks. Class M overlapped with the mineral occurrences and represents local anomalies.

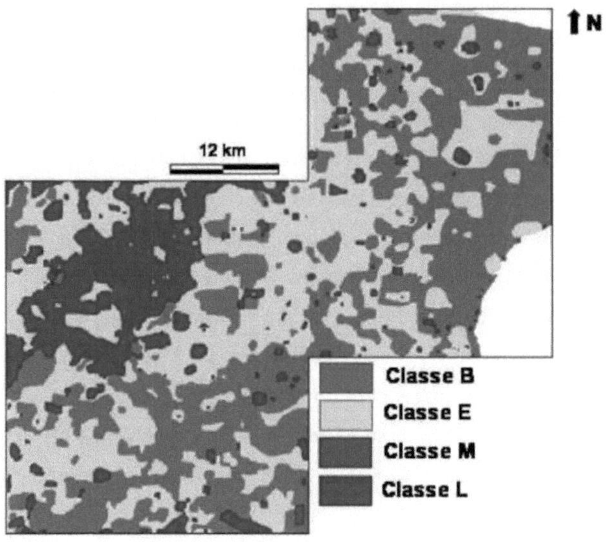

Figura 12. Result of digital processing of stream sediment geochemistry with orbital images in the MPVZ region, Uruguay, involving 14 variables: TM band 4 (vegetation), F factor and H factor, gamma radiometry, Ba, Co, Cu, Mn, Ni, Pb, P, V, Y and Zn (FILIPPINI-ALBA, 1998).

Regarding the geochemical background of current sediments in the area around *Isla Patrulla*, Uruguay, Filippini-Alba and Olivieira (1997) mentioned that the *Fraile Muerto* fault characterized two major domains: (a) *Pavas* and *Isla Patrulla* Formations; (b) *Lavalleja* Group. Domain (a) showed enrichment in Ba, Co, Cr, Ni and V compared to (b) and positive correlations between Fe, Mn, Co, Cr, Cu, Ni, V and Zn as well as contrasts with Ba and Pb. Factor analysis confirmed the multivariate correlation between phemophilic elements for factor 1 with 31% of the variance explained (Table 8). Factor 2, with 15% of the variance explained, showed the position of Ba and Pb versus Cr and Ni, probably derived from the alternation of felsic and mafic rocks, which characterizes the *Pavas* Formation. The Fe, Cu, V and Zn association (factor 3, 17%) was related to the phenomena of trace element adsorption on Fe oxides.

Table 8. Comparison of the results of the factor analysis in R mode for the (a) O and (b) L domains in the study area around the town of *Isla Patrulla*, Uruguay. *Loading* cut-off = 0.4.

Factor	Domain (a)		Domain (b)	
	Elements	Variance %	Elements	Variance %

1	Fe, Mn, Co, Cr, Cu, Ni, V	31	Fe, Mn, Co	19
2	Ba, -Cr, -Ni, Pb	15	Cu, V, Y	19
3	Fe, Cu, V, Zn	17	Cr, Ni, Zn	19
4	P	9	Ba, Pb, Zn	13
5	Y	9	P, -Ba, Zn	11

Domain (b) was enriched in P, B, Pb, Zn and slightly more in Fe and Cu than domain (a), but the correlations were not as well defined. Factor analysis distributed the explained variance more evenly. Factor 1 (Fe, Mn and Co, 19%) characterized adsorption phenomena on Fe and Mn minerals, which are known to have sedimentary genesis. Factors 2 (Cu, V and Y, 19%), 3 (Cr, Ni and Zn, 19%) and 4 (Ba, Pb and Zn, 13%) were associated with the volcanic, green schist and carbonate rocks that occur in the *Lavalleja* Group.

Local anomalies

The Mahalanobis distance is obtained by taking the square root of the result of pre- and post-multiplying the variance-covariance matrix of N variables by the difference vector between the value of each variable and its mean (GARRETT, 1989). This is a sensitive variable for detecting *outliers* and/or multivariate anomalies, which was superimposed on the interpolated map of factor 1 (Figure 11). In practice, the procedure was efficient for characterizing lithological anomalies related to the occurrence of mafic/ultramafic rocks and probably associated with mineralization, but it failed to detect the Pb anomaly on the Gutierrez Sheet. The Au mine was characterized by a main anomaly and a secondary anomaly, associated with the Mahalanobis distance and the other two occurrences were presented as regional anomalies, in the case of the amphibolites near *Zapicân* and the limestone near *Maria Albina*, in the latter case with low local contrast. It seems clear that when the color map is associated with the *Pavas* Formation (lilac to pink colors), the anomalies appear to be lithological in nature, with the anomalies with a possible metallogenetic association being related to the *Lavalleja* Group (yellow color).

Figure 13 represents a new attempt to contrast local anomalies with the geochemical *background* or with regional and lithological anomalies. The colored map is an improved version of the geochemical *background*, due to the participation of the total factors.

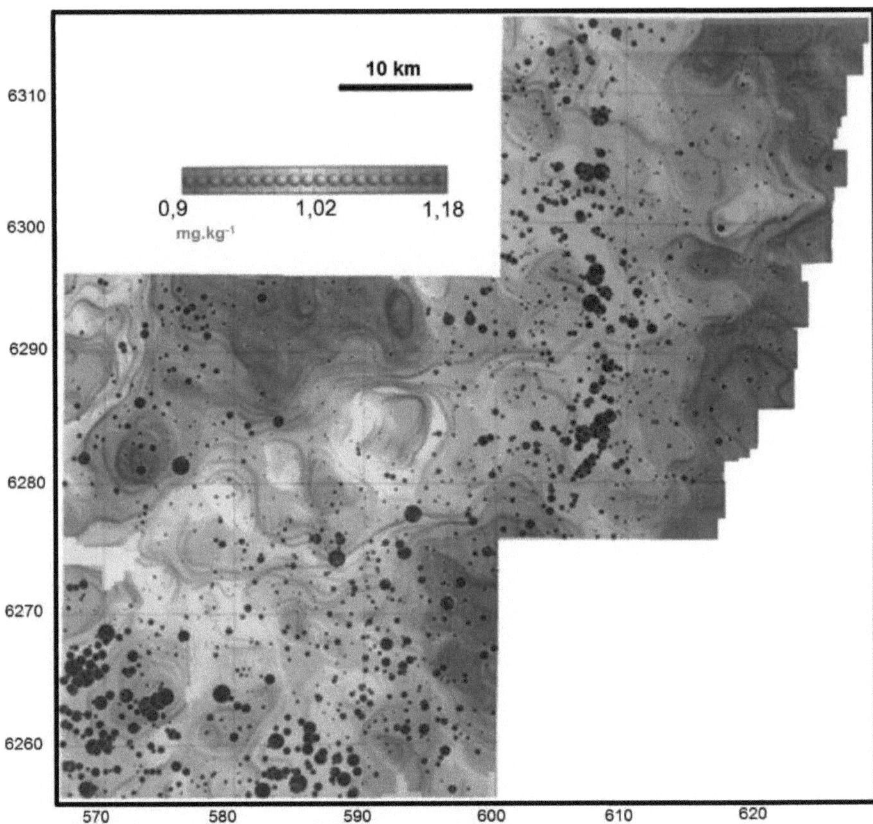

Figura 13. Circle map of Cu residuals (Real - Model) over the interpolated map of Cu modeled by linear regression as a function of factors (Table 6) in the MPVZ region, Uruguay.

Cu_{model}= 0.0846.F1 + 0.0125.F2 + 0.0166.F3 + 0.0072.F4 + 1.009; where F1, F2, F3 and F4 are the respective factors. Map produced by the Geological Survey of Finland.

Similar to the color map in Figure 11, *there is* a significant influence of factor 1 with a high percentage of explained variance (47%). The regression model explained 66% of the Cu variance. The map of circles (Cu residuals) significantly highlighted the *Barriga Negra* Group and the *Lavalleja* Group and erased the lithological anomalies relating to amphibolites. The occurrence of Au was notably contrasted, however, the other mineral occurrences were indicated by anomalies secondary to the residuals.

Figure 14 highlights the M group related to local anomalies and inserts the geochemical anomalies interpreted by statistical processing, mainly by integrating the circle maps for the base metals (Cu, Pb and Zn). The mineral occurrence of Pb in the *Gutiérrez* sheet was evidenced by a group M occurrence, as was the other Pb occurrence and the Au occurrence, however, the Fe-Cu occurrence appeared as class B but with great imprecision. All the lithological anomalies derived from the statistical treatment were positioned in the Pavas Formation, two of which were absent from the M group, but there was a significant occurrence of the L class. All the anomalies, lithological or not, but with the presence of the M class have potential for prospecting in later phases.

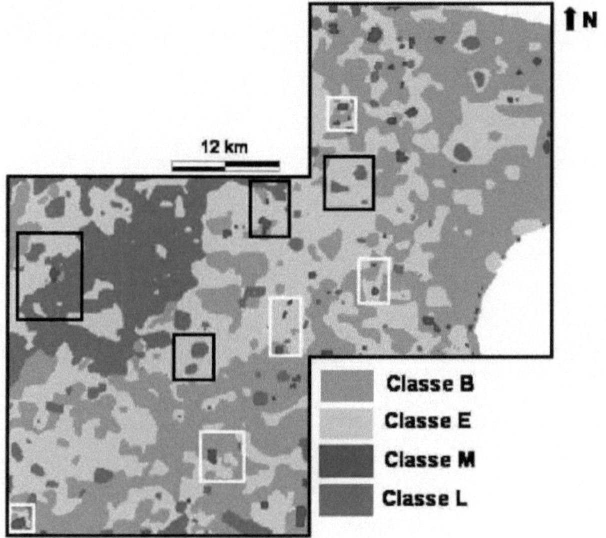

Figura 14. Result of unsupervised classification of 14 layers of information with M-class highlighting and overlay of lithological anomalies (dark rectangles) and those possibly related to mineral deposits (light rectangles).

Filippini-Alba (1998) established that the maps of overlapping Cu-Pb circles detected mineral occurrences in the MPVZ region with samples located at 0.6 to 1.6 km and an average of 1.2 km; while the map of unsupervised classification varied between 1.3 and 1.9 km with an average of 1.6 km. In any case, the interpretation potential of the latter map is more adequate, as it allows us to differentiate between lithological anomalies and those derived from mineralization.

Regarding the surroundings of Isla Patrulla, Filippini-Alba and Oliveira (1997) characterized three groups of anomalous current sediments, one of lithological origin and two associated with mineralization. In the lithological case, there are more than 30 samples with high levels of Co, Cr and Ni and moderate levels of Fe, Mn, Cu, V and Zn (S3), located on the *Isla Patrulla* Formation and the *Fraile Muerto* fault, suggesting that they are linked to the mafic and ultramafic rocks to the west of the fault. In the Lavalleja Group there are Fe, Cu, V and Zn anomalies (S4), which add Mn, P, B, Ba, Co and Pb to a second set (S5). This suggests the occurrence of anomalies derived from adsorption on Fe oxides and anomalies resulting from the adsorption of Fe - Mn oxides (relationship with the lithotypes of the *Lavalleja* group?).

Considering all the mineral occurrences already known, four in the MPVZ region and one in the vicinity of *Isla Patrulla*, in a total of five occurrences, only one of them was evidenced with difficulty, so the minimum efficiency of detecting mineral occurrences by the geochemistry of current sediments in the Uruguayan crystalline would be 80%.

CHAPTER 5

Geochemistry of quartz veins

As mentioned above, this chapter takes place in the study area located around *Isla Patrulla*, Uruguay, as described in Chapter 3.

Preliminary statistics for the population of 228 quartz veins showed the occurrence of extreme values for various elements (Table 9). In the case of Au, there were many samples below the detection limit (41% of samples), resulting in unreliable results.

The detection limits for As, B, Mo, Sb, Sn and W in current sediments were 20, 20, 2, 10, 20 and 10 $mg.kg^{-1}$ respectively. To try to assess the detection limit for quartz veins, the median value was tripled (Table 9), giving the values 9, 0, 6, 6, 0 and 0 $mg.kg^{-1}$ respectively. Thus, the detection limits for quartz filaments would be overestimated if the detection limits for current sediments were taken into account. The explanation for this would be the purity of the sample matrix in the case of veins. Thus, elements traditionally discarded in the statistical treatment of current sediments were used for the veins, mainly As, Mo and Sn.

Table 9. Preliminary statistics for the 228 quartz filament samples for 18 of the elements analyzed. SD = Standard Deviation; Q = Quartile, CV = Coefficient of Variation.

Variable	Minimum	Q25%	Median	Q75%	Maximum	DP	CV %
Fe_2O_3, %	0	1	2	2	16	1	80
Mn, $mg.kg^{-1}$	164	728	1130	1341	18070	1164	100
P, $mg.kg^{-1}$	15	32	48	79	664	73	110
Ag, $mg.kg^{-1}$	0	0	0	0	34	2	850
As, $mg.kg^{-1}$	0	0	3	6	204	20	330
Au*, $\mu g.kg^{-1}$	5	20	40	40	3200	222	370
B, $mg.kg^{-1}$	0	0	0	0	69	5	690
Ba, $mg.kg^{-1}$	0	7	10	19	618	54	260
Co, $mg.kg^{-1}$	0	1	2	4	230	15	380
Cr, $mg.kg^{-1}$	1	9	13	18	114	21	110
Cu, $mg.kg^{-1}$	12	18	22	34	764	64	160
Mo, $mg.kg^{-1}$	0	1	2	3	14	2	90
Pb, $mg.kg^{-1}$	0	2	4	6	15340	1025	1200
Sb, $mg.kg^{-1}$	0	0	2	3	14	2	90
Sn, $mg.kg^{-1}$	0	0	0	1	9	1	220
V, $mg.kg^{-1}$	0	1	2	4	76	9	180

W, $mg.kg^{-1}$	0	0	0	0	2990	198	1400
Y, $mg.kg^{-1}$	0	0	0	0	6	1	350
Zn, $mg.kg^{-1}$	1	4	6	10	740	51	420

* The samples were analyzed using three different methods at different institutions with lower detection limits of 5, 20 and 40 $\mu g.kg^{-1}$.

Deformed and asymmetrical distribution laws resulted from the Chi^2 and Kolgomorov-Smirnov tests, when only Ba and Mn fitted the theoretical distributions. The histograms for As, Cr, Cu, V and Zn suggested a tendency towards pluirimodality. The analysis of variance considering the Iitotypes of the quarries showed that the phyla related to the Lavalleja Group are depleted in Cr and enriched in Fe, P, Ba, Cu, Pb, V and Zn compared to the *Isla Patrulla* and *Pavas* Formations. In the first case, there was a multiple Fe-Mn-P-Ba-Co-Cr-Cu-Pb-V-Zn correlation, demonstrating the strong control of Fe (and Mn?) minerals over the geochemistry of the quartz veins. On the other hand, the formations located to the west of the *Fraile Muerto* fault showed a clear Fe-Mn-Cr correlation (FILIPPINI-ALBA, 1992).

Considering the preliminary analysis, *cluster* analysis and discriminant analysis, 11 *outliers* were separated and the 217 remaining quartz vein samples were classified into five groups (Figure 15), the description of which is shown in Table 10. The samples in group 1 represent the geochemical background, with the even groups being associated with Au mineralized, secondary and primary veins respectively. Groups 3 and 5 are linked to associations that suggest contamination of the crust with mafic deposits.

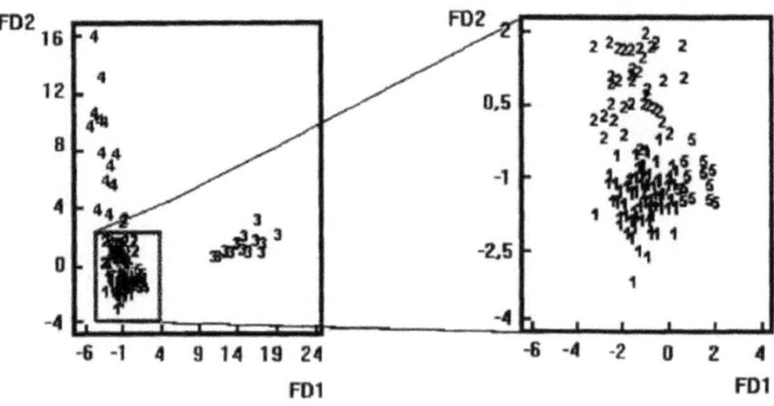

Figura 15. Dispersion diagram of the discriminant functions (DF, specified in continuation) and breakdown for groups 1, 2 and 5 defined for a population of 217 samples of quartz veins from the *Isla Patrulla* region, Uruguay.

FD1 = 0.102.Fe -0.00033.Mn - 0.0152.Au + 0.197.Cr + 0.00302.Pb - 0.0651.V + 0.00534.Zn - 2.77

FD2 = -0.753.Fe - 0.00003.Mn + 0.0433.Au + 0.040.Cr - 0.00832.Pb + 0.219.V - 0.0172.Zn - 1.49

Table 10. Characteristics of the groups of quartz vein samples from the Uruguayan Proterozoic (Isla Patrulla).

Group	NA	Geochemical relationship	Interpretation
F1	125	Low levels	*Background*
F2	50	Moderate enrichment in Fe, Mn, P, Au, Co, Mo and V. Zonality with group 4.	Secondary auriferous phyla.
F3	15	Significant enrichment in Fe, P and Cr and moderate in Co, Cu and V. Zonality?	Crustal contamination by mafic fills?
F4	14	Significant enrichment in Fe, P, Au, Co, Cu, Mo and V and moderate in Mn, As, Ba, Cr and Zn. Zonality with group 2.	Primary auriferous phyla.
F5	13	Moderate enrichment in Fe, Mn, P, Cr, Cu and Mo.	Associated with group 3?

NA = number of samples.

The samples in group 1 (*background)* were classified according to their location to the east (*Lavalleja* Group) or west (*Isla Patrulla* and *Pavas* Formations) of the shear zone (*Fraile Muerto* Fault), and a dispersion diagram was drawn up for the first two principal components (Figure 16). There is a well-characterized separation, and the PCs can be interpreted according to multielement associations:

CP1 → Fe, Ba, (Mn), P, Cu, V e Zn

CP2→ (Fe), Mn, Cr, -Cu e –Zn

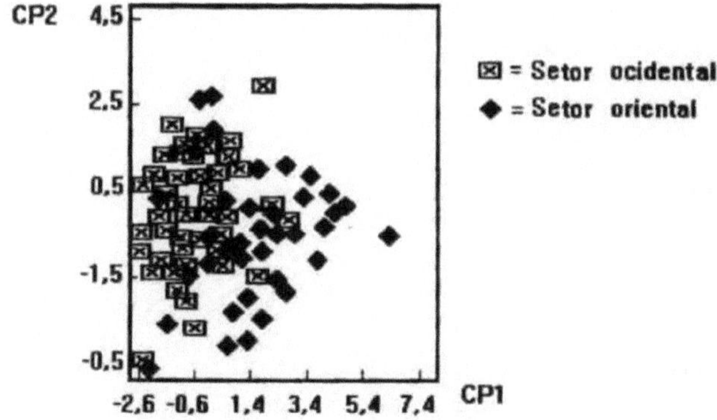

Figura 16. Dispersion diagram of CP1 and CP2 scores for quartz vein samples from the Isla Patrulla region, Uruguay. The PCA was applied to subgroup 1 with the data transformed by the square root.

CP1 = 0.48.Fe_2O_3+0.25.Ba+0.22.Mn+0.42.P+0.02.Cr+0.27.Cu+0.38.Pb+0.40.V+0.33.Zn (35%)

CP2 = 0.26.Fe_2O_3-0.01.Ba+0.58.Mn-0.12.P+0.64.Cr-0.32.Cu-0.10.Pb-0.03.V-0.25.Zn (18%)

The spatialization of the previous groups clearly illustrates a certain zonality, which can be seen in the north-central sector of the study area, with F2 around F4 and F5 nearby (Figure 16). In the south-central sector, F2 and F4 maintain the same configuration, but with F3 nearby.

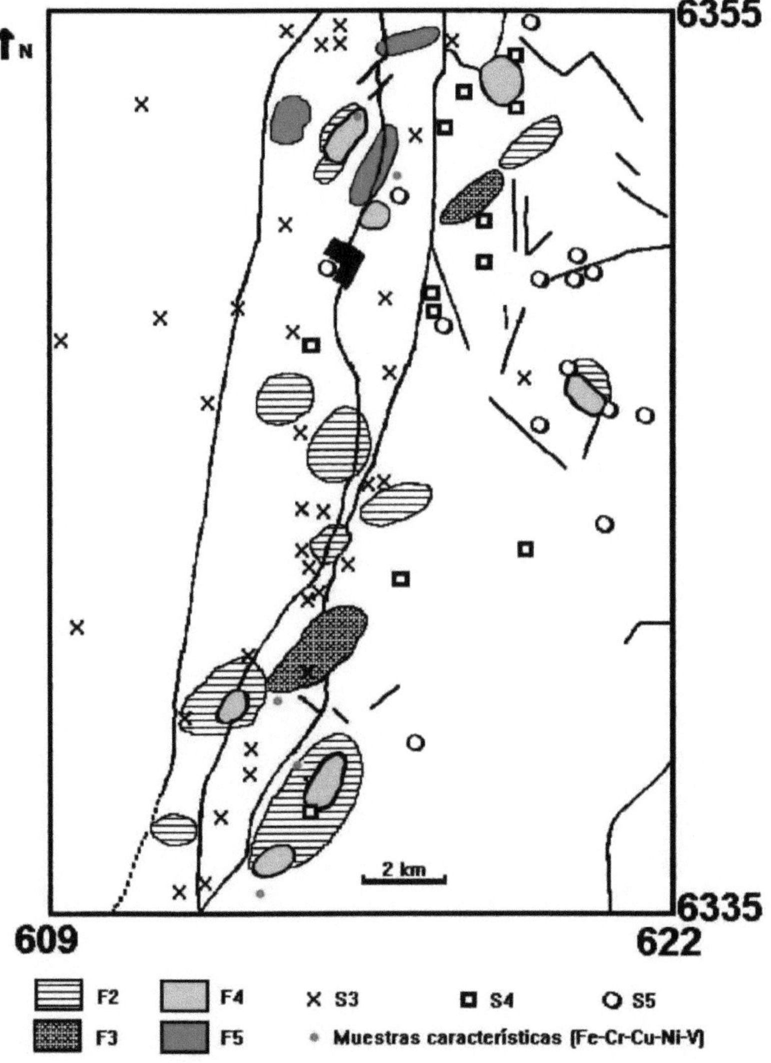

Figure 16. Comparison of anomalous geochemical groups for quartz veins and current sediments

around *Isla Patrulla*, Uruguay. F2, F3, F4 and F5 refer to the quartz vein groups and S3, S4 and S5 to the current sediment groups.

Au did not show any characteristic geochemical association, nor does it seem to be related to any specific mineral species. The analysis of the data suggested a local nature of the association of Au with femalophile or chalcophile elements, indicating its probable occurrence in quartz, sulphides, Fe oxides and possibly also in Mn oxides, minerals that controlled the geochemistry of the phylums. A lithological control by mafic-ultramafic or green schists is possible, but has not been clearly demonstrated. The Au associated with the *Isla Patrulla* formation would occur in its native form in quartz veins that would have mafic rocks as fillings; but in the case of the *Lavalleja* group, it would occur associated with sulphides, perhaps as a secondary element. The proximity of the *Fraile Muerto* fault to most of the Au occurrences suggests a structural control for the process.

Thanks

In institutional terms, I would like to thank BGR, CNPq, DAAD, DINAMIGE, GSF, UAÂ, UNICAMP and USP, who helped to make resources available, data, *software* and facilities available for the development of the process that led to this book.

From a personal point of view, a huge thank you to all the people mentioned in the preface, who, in addition to their valuable contribution, were witnesses to the effort and dedication involved in the process as a whole.

Bibliographical references

ADRIANO, D. **Trace elements in the terrestrial environment**. New York: Srpinger-Verlag, 1986, 533p.

AFIFI, A.A.; AZEN, S.P. **Statistical analysis**: A computer oriented approach. New York: Academic press, 1972. p. 228-230.

BONHAM-CARTER, G.F. **Geographic information systems for geoscientists**. Ottawa: Pergamon, 1994. 397 p.

BOSSI, J. Evidencias geológicas sobre la posible existencia de un nùcleo cratónico de más de 2000ma en el noroeste de Uruguay. In: Simpósio Sul-Brasileiro de Geologia, 3., Curitiba, 1987, **Atas**, v. 2, Curitiba, p. 821-831, 1987.

BOWELL, R.J.; ALPERS, Ch. A.; JAMIESON, H.E.; NORDSTOM, D.K.; MAJZLAN, J. The environmental geochemistry of Arsenic: An overview. **Reviews in Mineralogy & Geochemistry** v. 79, p. 1-16, 2014.

BOWIE, S.; THORNTON, I. **Environmental Geochemistry and Health**. Dordrecht: D. Reidel, 1985, 140p.

CHENG, Q.; AGTERBERG, F.P.; BONHAM-CARTER, G.F. A spatial analysis method for geochemical anomaly separation. **Journal of Geochemical Exploration** v. 56, p. 183-195, 1996.

CHORK, C.Y.; GOVETT, G.J.S. Interpretation of geochemical soil surveys by block averaging. **Journal of Geochemical Exploration**, v. 11, p. 53-71. 1979.

CRÓSTA, A.P. **Digital processing of remote sensing images**. Campinas: State University of Campinas, 1993. 170 p.

CRÓSTA, A.P.; MOORE, J. McM. Enhancement of landsat thematic mapper imagery for residual soil mapping in SW Minas Gerais state, Brazil: a prospecting case history in greenstonebelt terrain. In: THEMATIC CONFERENCE OF REMOTE SENSING FOR EXPLORATION GEOLOGY, 7, 1989, Calgary. **Proceedings...**, Calgary: ERIM, 1989. p. 1173-1187.

DARNLEY, A. International geochemical mapping: a new global project. **Journal of Geochemical Exploration** v. 39, p. 1-13, 1990.

DARNLEY, A.; BJORKLUND, A.; B0LVIKEN, B.; GUSTAVSSON, N.; KOVAL, P.; PLANT, J.;

STEENFELT, A.; TAUCHID, M.; XUEJING, X. **A global geochemical database for environmental and resource management**. Paris: UNESCO, 1995. 122 p.

DAVENPORT, P. Geochemical Mapping. **Journal of Geochemical Exploration**, v. 29, 212p., 1993.

DAVIS, J. **Statistics and data analysis in geology**. New York: Wiley, 1986, 645 p.

FESEFELDT, K.; VAZ, N.; ARRIGHETTI, R. Preliminary geotectonic interpretation of the Minas and neighboring areas. 1 map, 1988. In: FESEFELDT, K. **Unpublished report**. Hannover: BGR, 186p., 1988.

FIGUEIREDO, B.R.; CAPITANI, E.M. de; ANJOS, J.A.S.A. dos; LUIZ-SILVA, W.L. **Chumbo, ambiente e saùde**. Sao Paulo: ANNABLUME, 2012, 271p.

FILIPPINI-ALBA, J.M. **Application of statistical methods in geochemical prospecting of stream sediments and rocks in the Isla Patrulla region, Treinta y Tres district, Uruguay**. 1992. 178 p. Dissertation (Master's Degree in Geosciences) - Institute of Geosciences, University of Sao Paulo, Sao Paulo. 1992.

FILIPPINI-ALBA, J.M. **Analysis and integration of geochemical and remote sensing data in a sector of the Uruguayan Cristalino**. 1998. 172 p. Thesis (Doctorate in Science) - Institute of Geosciences, University of Sao Paulo, Sao Paulo. 1998.

FILIPPINI-ALBA, J.M.; CRÓSTA, A.P.; OLIVEIRA, S.M.B. Interpretation of surface geochemical data and integration with geological maps and Landsat - TM images for mineral exploration from a portion of the Precambriam of Uruguay. **Revista Brasileira de Geociências** v. 31, p. 123-130, 2001.

FILIPPINI-ALBA, J.M.; OLIVEIRA, S.M.B. Geochemical prospection of quartz veins and current sediments in precambrian terrains of Uruguay: statistical treatment of data. **Acta Geologica Leopoldensia** v. 20, n. 45, p. 25-45, 1997.

FILIPPINI-ALBA, J.M.; SOUZA FILHO, C.R. GIS-based Environmental Risk Analysis in the Ribeira Valley, Sao Paulo, Brazil. **Environmental Earth Sciences** v. 59, p. 1139-1147, 2010b.

GARRETT, R.G. The chi-square plot: a tool for multivariate outlier recognition. **Journal of Geochemical Exploration** v. 32, p. 319-341, 1989.

GÓMEZ-RIFAS, C. **The Sierra Ballena sinistral shear zone in Uruguay**. Sao Paulo: IG-USP, 244p., 1995 (Doctoral thesis, IG-USP).

GOVETT, G. **Rock Geochemistry in Mineral Exploration**. Amsterdam: Elsevier: 1983, 461 p.

HOWARTH, R. **Statistics and data analysis in geochemical prospecting**. Handbook of Geochemical Exploration 2. Amsterdam: Elsevier, 1983. 437 p.

JOLLIFFE, I. **Principal component analysis**. New York: Springer-Verlag, 1986. 271 p.

LÂG, J. Geomedical aspects of geochemical environments. In: PULKKINEN, E. (ed.) Environmental Geochemistry in Northern Europe. Geological Survey of Finland, **Special Paper** v. 9, p. 271-276,

1991.

LANCIANESE, V.; DINELLI, E. Different spatial methods in regional geochemical mapping at high density sample: An application on stream sediment of Romagna Apennines, Nothern Italy. **Journal of Geochemical Exploration** v. 154, p. 143-155, 2015.

MALCZEWSKI, J. GIS-based multicriteria decision analysis: a survey of the literature. **International Journal of Geographical Information Science** v. 20, n. 7, p. 703 - 726, 2006.

MASQUELIN, H. ; SILVA LARA, H.; BETUCCI, L.S.; DEMARCO, L.N.; PASCUAL, S.; MUZIO, R.; PEEL, E.; SCAGLIA, F. Lithologies, Structure and Basement-Cover Relationships in the Schist Belt of the Dom Feliciano Belt in Uruguay. **Brazilian Journal of Geology**, 2017 (in press).

NG, J.; WANG, J.; SHRAIM, A. A global problem caused by arsenic from natural sources. **Chemosphere** v.52, p. 1353-1359, 2003.

ROQUIN, C.; ZEEGERS, H. Improving anomaly selection by statistical estimation of background variations in regional geochemical prospecting. **Journal of Geochemical Exploration** v. 29, p. 295316, 1987.

PRECIOZZI, F.; SPOTURNO, J.; HEINZEN, W.; ROSSI, P. **Carta Geológica del Uruguay Scale 1:500.000**. Montevideo: DINAMIGE. 1 map, 1985.

PRECIOZZI, F.; SPOTURNO, J.; HEINZEN, W.; ROSSI, P. Memoria Explanativa de la Carta Geológica del Uruguay Escala 1:500.000. Montevideo: DINAMIGE. 92 p., 1988.

REICH, M.; LARGE, R.; DEDITIUS, A.P. New advances in trace element geochemistry of ore minerals and accessory phases. **Ore Geology Reviews** v. 81, p. 1215 - 1217, 2017.

ROLLINSON, H.R. **Using Geochemical Data**: Evaluation, presentation, interpretation. New York: Longman Sc. &Tech., 352p., 1993.

SHACKLETTE, H.; SAUER, H.; MIESH, A. Geochemical environments and cardiovascular mortality rates in Georgia. **U.S. Geological Survey Professional Paper** 574-C. Washington: USGS, 1970, 39p.

STANLEY, C.R.; SINCLAIR, A.J. Anomaly recognition for multi-element geochemical data. A background characterization approach. **Journal of Geochemical Exploration** v. 29, p. 333-353, 1987.

THORNTON, I. **Applied environmental geochemistry**. London: Academic Press, 1983. 501 p.

UNICAMP. **Medical Geology**. Metals, health and the environment. Campinas: UNICAMP, 2003 (International Workshop)

ZHOU, DI. Adjustment of geochemical background by robust multivariate statistics. **Journal of Geochemical Exploration** v. 24, p. 207-222, 1985.

Printed by Books on Demand GmbH, Norderstedt / Germany